A.P. 2864-C P.N.

ROYAL AIR FORCE PILOT'S NOTES

FOR

HORNET FIII

MERLIN 130 & 131 ENGINES

PREPARED BY DIRECTION OF THE MINISTER OF SUPPLY

A. Rowlands

PROMULGATED BY ORDER OF THE AIR COUNCIL

J. H. Barnes

"CROWN COPYRIGHT, REPRODUCED BY PERMISSION OF HER MAJESTY'S STATIONERY OFFICE"

©2012 Periscope Film LLC
All Rights Reserved
ISBN #978-1-937684-83-9
www.PeriscopeFilm.com

AMENDMENTS

Amendment lists will be issued as necessary and will be gummed for affixing to the inside back cover of these notes.

Each amendment list will, where applicable, be accompanied by gummed slips for sticking in the appropriate places in the text.

Incorporation of an amendment list must be certified by inserting date of incorporation and initials below.

A.L. NO.	INITIALS	DATE	A.L. NO.	INITIALS	DATE
1			7		
2			8		
3			9		
4			10		
5			11		
6			12		

This manual is sold for historic research purposes only, as an entertainment. It is not intended to be used as part of an actual flight training program. No book can substitute for flight training by an authorized instructor. The licensing of pilots is overseen by organizations and authorities such as the FAA and CAA. Operating an aircraft without the proper license is a federal crime.

NOTES TO USERS

THIS publication is divided into five parts: Descriptive, Handling, Operating Data, Emergencies, and Illustrations. Part I gives only a brief description of the controls with which the pilot should be acquainted.

These Notes are complementary to A.P.2095 Pilot's Notes General and assume a thorough knowledge of its contents. All pilots should be in possession of a copy of A.P.2095 (see A.M.O. A93/43).

Words in capital letters indicate the actual markings on the controls concerned.

Additional copies may be obtained by the Station Publications Officer by application on Form 294A, in duplicate, to Command headquarters for onward transmission to A.P.F.S., 81, Fulham Road, S.W.3 (see A.M.O. A1114/44). The number of this publication must be quoted in full—A.P.2864C—P.N.

Comments and suggestions should be forwarded through the usual channels to the Air Ministry (D.T.F.).

HORNET F111

AIR MINISTRY
December, 1947

A.P. 2864C—P.N.
Pilot's Notes

HORNET III

LIST OF CONTENTS

PART I—DESCRIPTIVE

Para.

INTRODUCTION ... ...

FUEL AND OIL SYSTEMS

Fuel tanks	1
Fuel booster pumps	2
Fuel cocks	3
Drop tanks	4
Fuel contents gauges	5
Fuel pressure warning lights	6
Priming system	7
Oil system	8

MAIN SERVICES

Hydraulic system	9
Pneumatic system	10
Electrical system	11

AIRCRAFT CONTROLS

Rudder pedals adjustment	12
Flying controls locking gear	13
Trimming tab controls	14
Undercarriage	15
Wing flaps	16
Wheel brakes	17

Para.

ENGINE CONTROLS

Throttle controls	18
Mixture control	19
Propeller controls	20
Engine cut-out controls	21
Supercharger control	22
Radiator flaps	23
Air intake filters	24

OPERATIONAL CONTROLS

Bomb, R.P. drop tank controls	25
Guns	26
Gyro gun sight	27

COCKPIT EQUIPMENT

Hood	28
Cockpit entrance	29
Seat adjustment	30
Heating and ventilation	31
Oxygen	32
Windscreen de icing	33

PART II—HANDLING

Management of the fuel system	34
Preliminaries	35
Starting the engines and warming up	36
Testing the engines and services	37
Taxying	38
Check list before take-off	39
Take-off	40
Climbing	41
General flying	42
Stalling	43

Diving	44
Aerobatics	45
Check list before landing	46
Approach and landing	47
Mislanding	48
After landing	49

PART III—OPERATING DATA

	Para.
Engine data—Merlin 130 and 131	50
Flying limitations	51
Maximum performance and range	52
Fuel consumptions	53

PART IV—EMERGENCIES

Engine failure during take-off	54
Engine failure in flight	55
Feathering	56
Unfeathering in flight	57
Single engine landing	58
Single engine overshoot	59
Glide landing	60
Undercarriage and flaps emergency operation	61
Fire-extinguishers	62
Hood jettisoning	63
Ditching	64
Emergency equipment	65

PART V—ILLUSTRATIONS

	Fig.
Instrument panel	1
Cockpit—left-hand side—seat removed	2
Cockpit—right-hand side—seat removed	3

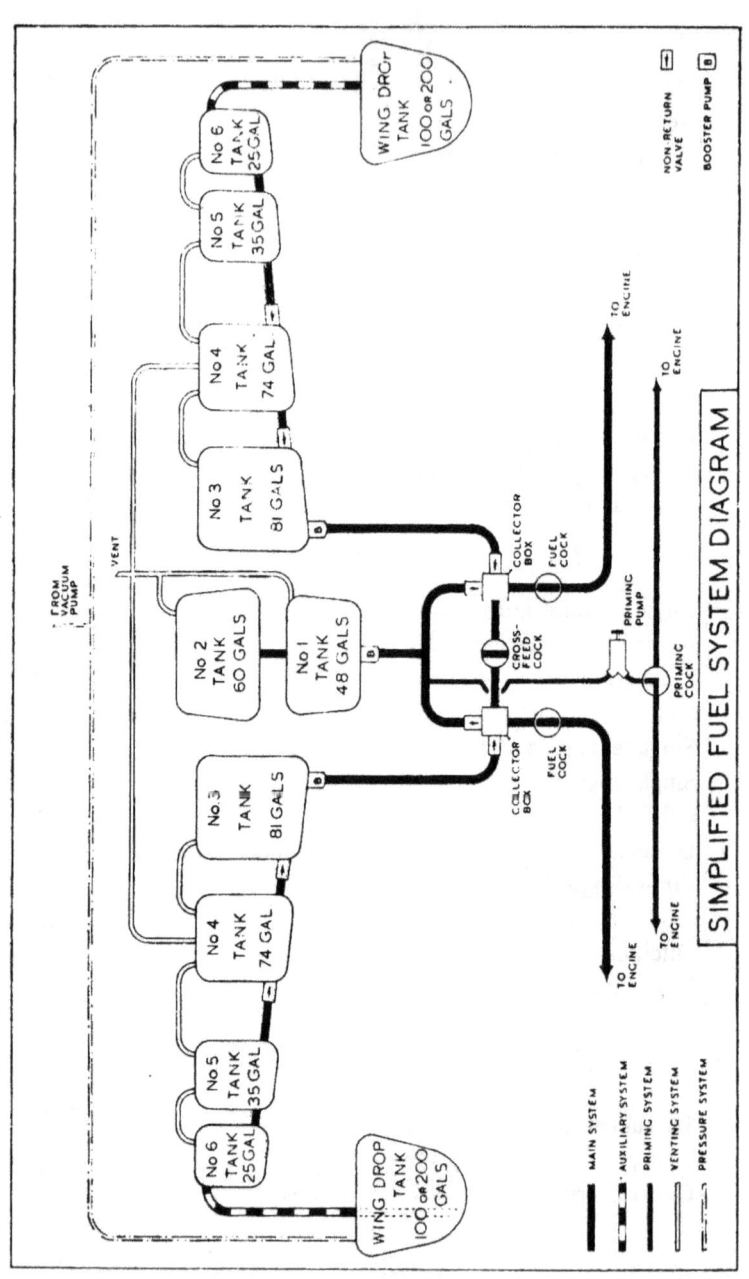

A.P. 2864C—P.N.
Pilot's Notes

PART I

DESCRIPTIVE

INTRODUCTION

The Hornet F Mark III is a mid-wing monoplane fighter, powered by one Merlin 130 on the port side, and one Merlin 131 on the starboard side, driving opposite rotating hydromatic four-bladed, fully-feathering propellers. Armament consists of four 20 mm. guns mounted in the nose, and there is provision for the alternative carriage of a bomb or a drop tank under each wing.

FUEL AND OIL SYSTEMS

1. **Fuel tanks**

    Fuel is carried in four pairs of wing tanks and two fuselage tanks. All these tanks are self-sealing and are vented to atmosphere. Each pair of wing tanks is interconnected, there being one pair of tanks outboard and one pair inboard of each engine. The outboard pair of tanks automatically feeds through a non-return valve into the inboard pair and thence to the respective engine, providing the fuel balance switch is CLOSED (see para. 3 [ii]). Early aircraft not embodying Mod. 245 have the two outboard pairs of wing tanks blanked off. The two fuselage tanks are interconnected, fuel draining from the upper to the lower tank and thence, by branching to both sides of the fuel balance cock, to both engines simultaneously. A drop tank can be carried under each wing, from which the fuel, which cannot be fed directly to the engines, is automatically transferred by air pressure from the exhaust side of the port engine vacuum pump to the respective out board pairs of wing tanks, which are thus continuously replenished until the drop tanks are emptied.

PART I—DESCRIPTIVE

The tank capacities are as follows:—

  Outer wing tanks:
    Port   25 + 35 = 60
    Starboard 25 + 35 = 60   120
  Inner wing tanks:
    Port   74 + 81 = 155
    Starboard 74 + 81 = 155   310
  Upper fuselage tank: 60
  Lower fuselage tank: 48

Total: 538 Galls.

Total with 2 × 100 gal. drop tanks   738 Galls.
Total with 2 × 200 gal. drop tanks   938 Galls.

NOTE.—The wing tanks are of flexible bag construction; the capacities vary, and particularly on new aircraft, before the bags have stretched, may be found to be somewhat less than the capacities quoted above.

2. **Fuel booster pumps**

Fuel is fed to the engines by three immersed fuel booster pumps, one in the bottom fuselage tank and one in each pair of inner wing tanks. They are controlled by switches (18) fitted below the centre of the instrument panel. In case of electrical or pump failure, fuel will be fed to the engines by gravity at a reduced flow, which may be insufficient for high power.

3. **Fuel cocks**

(i) There are two main fuel cocks (21) fitted on the front of the main spar, one on each side of the pilot's seat and they control the fuel flow to their respective engines. They are marked CLOSED—FUEL COCK—OPEN.

(ii) A cross-feed cock, mounted between the two pairs of wing tanks, is operated by an electro-pneumatic ram, controlled by a switch (17) labelled OPEN—FUEL BALANCE—CLOSED, on the right-hand side of the fuel booster pump switches. When this cock is CLOSED, each group of wing tanks feeds its respective engine. The fuselage tanks feed both engines. When this cock is OPEN

PART I—DESCRIPTIVE

all tanks feed both engines. When this cock is OPEN and the main fuel cock to one engine is CLOSED all tanks feed the other engine.

NOTE.—In the event of a complete electrical or pneumatic failure the cock will open automatically, irrespective of the setting of the switch.

4. **Drop tanks**

The drop tanks, which should only be jettisoned in cases of operational necessity, are released by the shielded pushbutton (46) mounted on a separate bracket aft of the electrical panel. For release speeds see para. 51 (ii).

5. **Fuel contents gauge**

A fuel contents gauge (16), incorporating three dials, records the contents of the lower fuselage tank and the port and starboard pairs of inboard wing tanks. The fuselage tank gauge will start to read when the level of the fuel has dropped to the lower tank. Below this gauge there is a further gauge which incorporates two dials and records the contents of the port and starboard pairs of outboard wing tanks. There are no fuel gauges for the drop tanks, but the gauge for the corresponding pair of outboard wing tanks (or inboard tanks on earlier aircraft without Mod. 245), will begin to show a fall in the fuel level when the drop tank has been emptied. The gauges do not read accurately in the ground attitude.

6. **Fuel pressure warning lights**

Two fuel pressure warning lamps (6) are mounted on the left-hand panel and indicate when the pressure in the fuel lines drops below 3 lb./sq. in.

7. **Priming system**

Two Ki-gass priming pumps (24) are fitted on the left-hand cockpit wall below the undercarriage selector lever. The forward pump serves the starboard engine. Later aircraft have one Ki-gass priming pump and a two-way selector cock. Priming fuel is drawn from the lower fuselage tank.

PART I—DESCRIPTIVE

8. Oil system

(i) Oil is supplied from two tanks, one in each wheel bay. Each tank has 12½ gallons oil capacity and 3 gallons air space.

(ii) There are no separate oil cooler controls as the radiator flaps also serve the oil coolers.

(iii) Two auxiliary oil tanks of 4½ gallons oil capacity each are situated in the port and starboard mainplane. They are used only when wing drop tanks are installed, and each auxiliary automatically replenishes the corresponding main tank.

(iv) An oil dilution system is fitted and is controlled by a pushbutton and two selector switches (35) (one for each engine) on the right-hand side of the instrument panel, below the ignition switches. For correct dilution periods see para. 49.

MAIN SERVICES

9. Hydraulic system

(i) Hydraulic pressure, supplied by a pump driven by the port engine, is stored in a pressure accumulator, from which it operates:—

Undercarriage and tail wheel.
Flaps

when the appropriate selector lever is placed in the required position.

(ii) Sufficient pressure will be available in the accumulator for one complete one-way operation of the undercarriage and flaps after feathering the port engine or failure of the engine-driven hydraulic pump.

(iii) A handpump (20) is provided on the left of the pilot's seat. With the appropriate selector lever in the required position, operating the handpump will transmit hydraulic fluid from a reserve supply in the reservoir direct to the jacks (without going through the accumulator) and under sufficient pressure to operate the desired service at a reduced rate. If Mod. 339 is incorporated the handpump operates the flaps only unless a cock, operable on

PART I—DESCRIPTIVE

the ground only, is first turned on; the handpump can then be used to operate the flaps or undercarriage for ground testing purposes.

(iv) An emergency compressed air system is provided. With the appropriate selector in the desired position, the red painted toggle lever (19) on the left and to the rear of the pilot's seat must be pressed, given a quarter turn anti-clockwise and pulled. The undercarriage or flaps will then be lowered through the normal pipelines by air pressure instead of hydraulic pressure (but see para. 16 (iv)). Once they have been lowered by air pressure they cannot again be raised. Before the hydraulic system can again be used it must be vented of all air. If Mod. 339 is incorporated the flaps cannot be lowered by the compressed air emergency system.

> NOTE.—Some early aircraft have the emergency lever situated on the front spar to the left-hand side of the cockpit. Operation is the same as above.
> The emergency compressed air container is situated aft of the ammunition boxes in the fuselage decking. A gauge in the container shows the available pressure which should read 1,800 lb./sq. in. max. 1,600 lb./sq. in. min.

10. Pneumatic system

(i) A compressor on the starboard engine charges a bottle for the operation of:—

    Brakes

    Electro-pneumatic rams for:

        Air intake filters
        Automatic supercharger gear change
        Fuel cross-feed cock
        Radiator flaps.

The available pressure is shown on the triple pressure gauge (30) on the panel forward of the engine control box; this should record 320 lb./sq. in. under normal conditions. All services except the brakes are cut off by a pressure maintaining valve if the pressure falls below 150 lb./sq. in.

PART I—DESCRIPTIVE

(ii) Two vacuum pumps, one on each engine, together operate the instrument flying panel. If either pump should fail it is automatically isolated from the system. Each pump can be proved separately on the ground by alternately starting the port and starboard engine first, and checking that the artificial horizon erects properly.

11. **Electrical system**

(i) Two generators, one on each engine, charge a battery and supply electrical power at 24 volts for the operation of:—
Bomb selection, fusing and release
Camera gun
Electro-pneumatic rams for:
    Air intake filters
    Supercharger gear change
    Fuel cross-feed cock
    Radiator flaps
Engine starters and booster coils
Feathering pump motors
Fire-extinguishers
Flaps position indicator
Fuel booster pumps
Fuel pressure warning lights
Guns
Gyro gun sight
Instrument panel and U.V. lighting
Oil dilution
Pressure-head heater
Radio
Recognition, identification, and landing lights
R.I. compass
Undercarriage warning lights
Wing drop tank jettison mechanism

NOTE.—Some early aircraft not embodying Mod. 338, have only one generator on the port engine.

(ii) Two generator warning lights are fitted on the top right-hand side of the instrument panel. A ground/flight switch (45) is fitted on the right-hand cockpit wall level with the pilot's elbow. When the switch is at **FLIGHT** the generator warning light comes on and remains on when the

PART I—DESCRIPTIVE

A.L.1
PART I
para. 11 (ii)
page 13

generators are not charging the battery. The ground/flight switch only isolates the battery from the electrical services. When either engine is running, its generator will operate all these services even if the ground/flight switch is at GROUND. The electrical services can, therefore, only be isolated while the generators are charging if they are switched OFF individually.

| The perspex cover on the electrical panel is held in the up position by a metal lug when the ground/flight switch is at FLIGHT.

(iii) A ground starter socket is fitted in the rear of the starboard wheel well.

AIRCRAFT CONTROLS

12. Rudder pedals adjustment

The rudder pedals can be adjusted for reach by the toes.

13. Flying controls locking gear

The rudder pedals are locked by a tubular bar which is connected to the control column locking device by a cable. The locking gear is stowed in the fuselage decking underneath the sliding hood.

14. Trimming tab controls

The elevator trimming tab control (32) is on the left of the pilot's seat, and the indicator is on the panel, forward of, and below, the engine control box. The rudder trimming tab control (33) and indicator are mounted on a pedestal aft of the elevator trimming tab control. The aileron trimming tab control and indicator (39) (if fitted) operates the starboard aileron only, and is on the right-hand side of the instrument panel.

15. Undercarriage

(i) The undercarriage and tail wheel selector lever (25) is on the rear face of the engine control box and has two positions only: UP—DOWN. There is a spring-loaded catch under the knob of the lever which must be squeezed before the selector can be moved. As long as the wheels are on the ground the lever is locked in the DOWN position by a solenoid. The solenoid can be overridden, to permit undercarriage UP to be selected on the ground in case of emergency, by a switch below the selector lever.

PART I—DESCRIPTIVE

(ii) The undercarriage position indicator (8) is fitted on the top left-hand side of the instrument panel and has dimmer screens for night flying. Indications are as follows:—

Wheels locked UP—No lights

Wheels between UP and DOWN—Two red lights

Wheels locked DOWN—Two green lights

There is no warning horn, but an additional red light (8) alongside the main indicator lights up if the wheels are not locked DOWN when the throttles are less than one quarter open. No tail wheel indicator is fitted.

(iii) If engine-driven pump or accumulator pressure is not available, the undercarriage and tail wheel can be lowered, at a reduced rate, by selecting DOWN and operating the handpump. About 120 double strokes will be required.

(iv) If the undercarriage cannot be lowered by hydraulic pressure, it can be lowered by the emergency compressed air system by operating the red painted toggle lever (19) on the left of the pilot's seat. The normal selector must be in the DOWN position. (On some early aircraft a red painted cock on the forward face of the front spar, on the pilot's left, is fitted instead of the toggle lever). Once the undercarriage has been lowered by compressed air, it cannot again be raised.

16. Wing flaps

(i) On aircraft embodying Mod. 44, operation of the wing flaps is controlled by the selector lever (26) marked FLAPS on the engine control box. It has three positions and by putting the flap lever at the middle notch on the quadrant, the maximum lift position, approximately 25°, is automatically obtained. With the lever fully down the maximum flap angle obtained is 75°. There is a tendency for the flaps to creep when the selector is in any position other than fully UP, max. lift, or DOWN.

(ii) The flaps position indicator (9) is mounted on the left-hand side of the instrument panel inboard of the undercarriage position indicator. The max. lift position is shown by a red mark.

PART I—DESCRIPTIVE

(iii) Even if the flaps cannot be lowered by the handpump (which will take about 30 double strokes) it is recommended that the flaps should not be lowered by the compressed air system. (If Mod. 339 is incorporated they cannot be). Pressure may be insufficient to lower them equally or fully, or to keep them down if speed is increased, nor can they be raised again in the case of mislanding. Further there may be insufficient air pressure to lower the undercarriage as well.

(iv) On early aircraft not embodying Mod. 44, the flap selector also has three positions but in this case they are UP, neutral and DOWN. The lever should be left at UP in flight. The maximum lift or other intermediate positions can be selected by setting the lever to DOWN and then to the neutral position when the desired flap angle has been reached. There is a tendency for the flaps to creep when the selector is not set to UP or DOWN.

17. Wheel brakes

The brake control lever and parking catch are on the control column. Differential braking is obtained by operating the rudder pedals. In the event of failure of the compressor, brake pressure will still be available when the supply falls below 150 lb./sq. in. although all other pneumatic services will then become inoperative. A triple pressure gauge, showing the air pressure in the pneumatic system, and at each brake, is mounted on the panel, forward of the engine control box. With the rudder bar central and brakes fully applied the pressure at each brake gauge should be 90 to 100 lb./sq. in.

ENGINE CONTROLS

18. Throttle controls

The throttle levers (28) are held by stops at the take-off position which give +18 lb./sq. in. boost. When the spring catches on the levers are pulled back, the levers can be moved fully forward to the combat boost position. The friction control for the throttle levers is the small black knob on the engine control box.

PART I—DESCRIPTIVE

A.L.1
PART I
para. 19

19. Mixture

There is no manual mixture control fitted as the mixture strength is automatically controlled by the S.U. injector pumps. Economical cruising is obtained at or below +9 lb./sq. in. boost and 2,650 r.p.m.

20. Propeller controls

The propeller control levers (29) for the hydromatic propellers vary the governed r.p.m. from 3,000 to 1,800. The propeller friction control is the white knob on the engine control box. Spring catches on the levers must be pulled back and the levers themselves moved into the feathering gate before the feathering pushbuttons will be operative. The feathering pushbuttons (10) are positioned to either side of the gyro gunsight at the top of the instrument panel, and there is a fire warning light inset in each button.

21. Engine cut-out controls

The slow running cut-out controls are mounted on the front face of the front spar, one on either side of the pilot's seat. They are spring-loaded and push inwards to stop the engines.

22. Supercharger control

(i) A two position switch (27) marked MOD—AUTO is fitted on the engine control box. The supercharger gear change is electro-pneumatically operated by an aneroid which is influenced by the speed as well as the height of the aircraft to compensate for ram effect. When the switch is set to MOD, the two-stage, two-speed superchargers remain in low gear at all altitudes. With the switch set to AUTO high gear will be engaged automatically, at approx. 8,000 feet which is the change height for a full-power climb only. However, when climbing at normal power (2,850 r.p.m., +12 lb. boost) a better rate of climb is obtained by delaying the gear change until the boost has dropped to +9 lb./sq. in. In this case climb with the switch at MOD and move it to AUTO when the boost has fallen to this figure.

PART I—DESCRIPTIVE

(ii) Operation of the gear change can be checked on the ground by pressing the test pushbutton mounted in each engine nacelle, on the bottom of the fireproof bulkhead. With the cockpit switch in the AUTO position depressing either button will cause both engines to change to high gear.

> NOTE.—In the event of failure of the electrical or pneumatic systems the supercharger will stay in, or return to, low gear at all altitudes.

23. Radiator flaps

The electro-pneumatic rams for the radiator flaps are operated by switches (41) at the forward end of the pilot's electrical panel. There are three positions, marked AUTO, MAN. OPEN, and EM'CY PRESS & TURN. With the switch in the AUTO position an inching unit operates the ram, and starts to open when the coolant reaches a temperature of 105°C. With the switch in the MAN. OPEN position the radiator flap is opened to its maximum extent. The EMERGENCY PRESS & TURN position is used only in the event of pneumatic or electrical failure. The stud in the top of the switch must be depressed before the switch can be moved to this position, and the radiator flaps then remain half open.

> NOTE.—In the event of pneumatic or electrical failure in flight, the rams become inoperative, and the radiator flap may then be closed by the slipstream. If excessive temperatures are experienced, the EMERGENCY PRESS & TURN position should be selected.

24. Air intake filters

There are two pushbuttons (42) marked FILTER IN—FILTER OUT on the electrical panel, to the right of the pilot's seat. The filters are electro-pneumatically operated. Extending the undercarriage will automatically bring in the filters (unless already selected) and they will stay in when the undercarriage is retracted. Therefore, in flight, the "straight through" airflow must be selected by depressing the FILTER OUT pushbutton. The filters can be used in flight, if required, by depressing the FILTER IN pushbutton.

PART I—DESCRIPTIVE

NOTE.—In the event of failure of electrical or pneumatic system, the filters will remain in the position selected at the time of the failure.

OPERATIONAL CONTROLS

25. **Bomb, R.P. and drop tank controls**

(i) On some aircraft a MASTER SWITCH, GUNS/R.P. selector, and a PAIRS or SALVO switches are fitted. These, however, are inoperative.

(ii) The shielded jettison pushbutton (46) for the wing drop tanks is mounted on a separate bracket aft of the pilot's electrical panel.

26. **Guns**

(i) The thumb operated switch on top of the control column is the firing control unit for the guns; it permits the port or starboard pair of guns to be fired separately or together. The flap over the switch is the safety control and must be raised before the guns can be fired.

(ii) The cine-camera master switch on the pilot's electrical panel must be ON before the pushbutton on top of the control column will operate the camera. The camera will operate automatically when the guns are fired if the camera master switch is ON.

27. **Gyro gun sight**

The sight is mounted above the instrument panel. The selector (43) is on the right-hand cockpit wall and the master switch is on the electrical panel. The ranging device is incorporated in the starboard throttle lever.

COCKPIT EQUIPMENT

28. **Hood**

(i) The sliding hood is opened and closed by the crank handle (40) mounted on the right-hand cockpit wall, and is automatically locked in any intermediate position when the handle is released.

PART I—DESCRIPTIVE

**A.L.1
PART I
para. 28 (ii)**

(ii) In emergency the hood may be jettisoned by means of the lever (13) marked HOOD JETTISON on the right-hand cockpit wall forward of the hood crank handle.

NOTE.—On some early aircraft not embodying Mod. 231, it is necessary first to wind the crank handle half a turn backwards before the hood can be jettisoned.

(iii) A pushbutton on the outside of the fuselage forward of the windscreen on the starboard side is pressed to permit the hood to be opened from outside.

29. Cockpit entrance

Entry is gained by a collapsible metal ladder which is hooked over the lower lip of the radiator ducting. The ladder is stowed behind the pilot's seat.

30. Seat adjustment

A lever on the right-hand side of the seat provides adjustment for height. The seat can also be adjusted for tilt by pulling up and releasing the toggle handle on the front of the seat.

31. Heating and ventilation

(i) The heating control is a small handwheel (22) on the left-hand cockpit wall, aft of the Ki-gass priming pumps. It is turned backwards to permit hot air from the port radiator to enter the cockpit.

(ii) Two ventilators are provided, one on the right-hand cockpit wall, forward of the hood crank handle, and one (7) on the left-hand cockpit wall beneath the coaming.

32. Oxygen

A Mk. XIA oxygen regulator flow indicator and high-pressure control (14) are mounted together on the right-hand side of the instrument panel.

33. Windscreen de-icing

A handpump with a regulator (15) is mounted to the right of the oxygen regulator panel.

A.P. 2864C—P.N.
Pilot's Notes

PART II
HANDLING

34. Management of the fuel system

(i) The booster pumps must never be switched on unless the engines are running, or fuel may be injected into the supercharger irrespective of the position of the slow running cut-out controls.

(ii) Take off with all booster pumps ON and the FUEL BALANCE switch CLOSED. At 2,000 feet, switch OFF the wing tank booster pumps. Fuel will then be drawn from the fuselage tank, and stability will be improved as this empties. When the fuselage tank is empty its booster pump should be switched OFF and the wing tank booster pumps switched ON.

NOTE.—There are no fuel gauges for the drop tanks, but the gauge for the corresponding pair of outboard wing tanks (or inboard tanks on aircraft without Mod. 245) will begin to show a fall in the fuel level when the drop tank has been emptied.

(iii) In the event of failure of one engine, the fuel from the failed engine wing tanks may be used to feed the live engine as follows:—
 (a) Set the failed engine fuel cock to CLOSED.
 (b) Set the fuel balance switch to OPEN.
 (c) Switch ON the booster pump in the tanks in the failed engine wing, and switch OFF the other two booster pumps.

A.L.1
PART II
para. 34 (iii)
(d)

 (d) When the failed wing tanks have been emptied, the balance switch should be set to CLOSED; the booster pumps for the live engine wing tank switched ON, and the booster pump for the failed engine wing tanks switched OFF.

35. Preliminaries

Before entering the cockpit check:—

The reading of the emergency air pressure gauge situated in the fuselage decking, which should be 1,600 to 1,800 lb./sq. in.

PART II—HANDLING

On entering the cockpit check:—

The correct operation of the flying controls and equal adjustment of the rudder pedals.

Ignition switches	OFF
Ground/Flight switch	GROUND
Booster pumps	OFF
Undercarriage emergency override switch ...	OFF
Undercarriage compressed air emergency control	Closed
Pneumatic supply pressure ...	150 lb./sq. in. min.

Then switch the Ground/Flight switch to **FLIGHT** and check:—

Undercarriage lever down and indicator lights green

Fuel gauges	Check contents
Flaps	Fully up (selector up)

36. Starting the engines and warming up

(i) Set the controls as follows:—

Fuel cocks	ON
Throttles	1 inch open
Propeller control levers	Fully forward
Supercharger switch	AUTO
Radiator flap switches	MAN. OPEN
Ground/Flight switch	GROUND

(ii) Operate the priming pump for the engine to be started until the pipelines are full. This may be judged by a sudden increase in resistance. Then prime the engine with the following number of strokes if it is cold:—

Air temp. °C.	+30	+20	+10	0
Number of strokes	3	4	7	12

(iii) Have a ground starter battery plugged in (24 volt), and switch on the ignition. Press the booster-coil and starter pushbuttons simultaneously and hold the booster-coil pushbutton depressed until the engine is running smoothly, continuing to prime if required. Screw down the priming pumps after use.

(iv) When both engines are running satisfactorily have the

PART II—HANDLING

ground starter battery disconnected and switch the Ground/Flight switch to FLIGHT.

(v) Open the throttles to 1,200 r.p.m. and warm up at this speed.

(vi) Switch on the R.I. Compass.

37. Testing the engines and services

While warming up.

(i) Check all temperatures and pressures. Test the operation of the hydraulic system by lowering and raising the flaps. Test each magneto in turn as a precautionary check before increasing power further.

After warming up to 15°C. (oil) and 40°C. (coolant). For each engine in turn :—

(ii) Open the throttle to zero boost, check the operation of the two-speed two-stage supercharger gear change by setting the cockpit switch to AUTO and having the ground crew press the testing pushbutton for 30 seconds. Boost should rise slightly and the r.p.m. flicker when high gear is engaged. Then change back to low gear (see para. 22 (ii)).

(iii) At the same boost thoroughly exercise the constant-speed propeller by moving the control lever over the full governing range at least twice, and return the lever fully forward.

Check that the generator warning light for each engine is out. (On early aircraft the port engine only).

(iv) At the same boost, and when the engine has cleared, check that the r.p.m. are within 50 of the reference figure, then check each magneto in turn. The single ignition drop should not exceed 150 r.p.m. It the drop exceeds this figure and rough running is not excessive it may be possible to clear the engine by carrying out the drill in sub. paras. (v) and (vi) below.

A.L.1
PART II.
para. 37 (iv)
NOTE

NOTE : The following comprehensive checks should be carried out after repair, inspection other than daily, if required by sub. para. (iv) above, or at the discretion of the pilot. Normally if the check in (iv) has been satisfactory NO USEFUL PURPOSE will be served by a full power check.

PART II—HANDLING

(v) With the propeller control lever fully forward, open the throttle to the gate and check take-off boost and static r.p.m. This check should be as brief as possible.

(vi) Throttle back to +9 lb./sq. in. boost, or further if necessary to ensure that r.p.m. fall below 3,000, and test each magneto in turn. The single ignition drop should not exceed 150 r.p.m.

> NOTE.—The engines warm up very quickly on the ground and it is strongly recommended that the aircraft be faced into wind during all ground running which should be kept to the minimum compatible with safe engine checking.

38. Taxying

(i) Before taxying, check the pneumatic supply pressure, which should be 320 lb./sq. in. If lower, ensure that it has built up since warming and running up the starboard engine.

(ii) The engines must not be allowed to idle below 1,000 r.p.m. for longer than necessary.

39. Check list before take-off

	At typical service load less external stores. (17,880 lb.)	At typical service load plus 2 × 100 gal. drop tanks. (19,550 lb.)
T—Trimming tabs		
Elevator	½ div. nose down	*(To be issued by amendment)*
Rudder	Neutral	
Aileron	Neutral	
P—Propeller levers	Fully forward	
F—Fuel	Main cocks OPEN. Check contents. All booster pumps ON. Fuel balance switch CLOSED.	
F—Flaps	UP (Selector UP). Maximum lift position for shortest run.	
Radiator flap switches	AUTO	
Superchargers	MOD (AUTO for combat)	
Sliding hood	Closed.	

PART II—HANDLING

40. Take-off

(i) If taxying has been prolonged, it is advisable to clear the engines before take-off.

(ii) Align the aircraft carefully on the runway, making certain that the tail wheel is straight.

(iii) Tighten up the throttle friction controls and open the throttles gently, keeping straight by coarse use of the rudder. There is no tendency to swing if the throttles are opened evenly, but the travel of the levers is small for the power obtained, and care is necessary to avoid a large difference in the boost pressure of the two engines.

(iv) The tail comes up easily and the aircraft can be pulled off the ground at about 110 knots (130 m.p.h.) I.A.S.

(v) When comfortably airborne brake the wheels, then retract the undercarriage.

(vi) (a) At typical service load with the flaps up, wheels up or down, safety speed at +12 lb./sq. in. boost is 135 knots (155 m.p.h.) I.A.S. Safety speed at +18 lb. boost is 145 knots (165 m.p.h.) I.A.S.

(b) At typical service load plus 2 × 100 gallon drop tanks, flaps up, wheels up or down, safety speed at +12 lb. boost is   knots (m.p.h.) I.A.S. Safety speed at +18 lb. boost is   knots (m.p.h.) I.A.S.

(Figures to be issued by amendment)

(vii) The aircraft accelerates rapidly to safety speed. If full take-off boost and flaps have been used, reduce to climbing power and raise the flaps when safety speed has been reached.

The aircraft has a good take-off at +12 lb./sq. in. boost and 3,000 r.p.m. and it is not normally necessary to exceed this power setting.

(viii) Unless operating in sandy or dust laden conditions depress the FILTER OUT pushbutton at 1,000 feet. Switch OFF the two wing-tank booster pumps.

PART II—HANDLING

41. Climbing

The recommended climbing speed is 175 knots (200 m.p.h.) I.A.S.

On a normal rated climb leave the supercharger switch at MOD and change to AUTO when the boost has dropped to +9 lb./sq. in.

42. General flying

(i) Stability.

At typical service load longitudinal and directional stability is satisfactory, but the aircraft is unstable laterally in all conditions of flight.

(ii) Changes of trim:

Undercarriage up ...	nose up
„ down ...	nose down
Flaps up	nose down
„ down	nose up
Radiator flaps open ...	strong and *delayed* nose up
„ „ auto ...	strong and *delayed* nose down

With the radiator flaps in the AUTO position adequate cooling is available for all conditions of flight. The radiator flaps should not be opened at high speed as with increased speed the change of trim induced is increased proportionately. Considerable buffeting will be felt if the radiator flaps are left open.

(iii) Controls

All controls are light and effective throughout most of the speed range, but the spring tab ailerons tend to lose effectiveness at low speeds.

The spring tab rudder control tends to stiffen at high speeds. The elevator and rudder trimming tabs are powerful and sensitive and must always be used with care. The aileron trimming tab (if fitted) is less effective.

PART II—HANDLING

(iv) Propellers

The four-bladed hydromatic propellers tend to overspeed when power is increased, during take-off and in dives. The propeller control levers must always be moved slowly and carefully and rapid throttle opening should be avoided. Synchronising engine speeds in the air may require constant manipulation of the propeller control levers.

(v) Flying at reduced airspeed.

Set the propeller control levers to give 2,650 r.p.m. and reduce speed to 155 knots (180 m.p.h.) I.A.S. Lower the flaps to max. lift and fly at about this speed.

43. Stalling

(i) The stalling speeds in knots (m.p.h.) I.A.S., engines off, are:—

	At typical service load (no external stores) 17,880 lb.	At typical service load, plus 2 × 100 gallon drop tanks, 19,550 lb.
Undercarriage and flaps up	118 (136)	*(To be issued by amendment)*
Undercarriage and flaps down	93 (105)	

(ii) Warning of the approach of the stall, which is not very well defined, is given by elevator buffeting which commences some 10 knots before the stall with the undercarriage and flaps up or down. At the stall the nose drops gently and recovery is straightforward and easy, but care must be taken to avoid pulling the control column back at all harshly after unstalling or the aircraft will stall again. Stalling speeds are not appreciably affected by the position of the radiator flaps, but buffeting is increased with them open.

PART II—HANDLING

(iii) Ample warning of the approach of the stall in a steep turn is given by pronounced elevator buffeting. The aircraft shows no tendency to flick out of the turn and recovery is immediate, if the backward pressure on the control column is relaxed.

44. Diving

(i) A considerable push force is necessary to hold the aircraft in a dive to the limiting speed; it should, therefore, be trimmed into the dive.

(ii) The aircraft is very clean and gains speed rapidly. Care is, therefore, necessary to avoid exceeding the speed/height limitations. (See para. 51 (i) .)

(iii) The aircraft is steady in the dive but care should be taken in the recovery to avoid large accelerations which might overstress the aircraft structure.

45. Aerobatics

(i) A large amount of height may be gained or lost in certain manoeuvres and an ample margin must always be allowed for recovery from aerobatics in the looping plane.

(ii) No attempt must be made to maintain inverted flight as oil pressure very quickly drops to zero.

(iii) The recommended initial speeds are:

Roll	230—250 knots (270—290 m.p.h.) I.A.S.
Loop	330—350 knots (380—400 m.p.h.) I.A.S.
Half roll off loop ...	350—365 knots (400—420 m.p.h.) I.A.S.
Upward roll	350 *plus* knots (400 *plus* m.p.h.) I.A.S.

NOTE.—Aerobatics are prohibited when carrying external stores.

A.L.1
PART II
para. 46

46. Check list before landing

On entering the circuit reduce speed to 175 knots (200 m.p.h.) I.A.S. and check :—

Pneumatic supply pressure	320 lb./sq. in.
Brake pressure at wheels	90–100 lb./sq. in.
Radiator flaps switches	AUTO
Fuel	Booster pumps ON.

Then lower the undercarriage and check indicator lights green and warning light out.
Reduce speed to 155 knots (180 m.p.h.) I.A.S. Lower the flaps to the max. lift position. For the final approach select 2,850 r.p.m. and full flap (leaving the selector fully down).

47. Approach and landing

At the maximum landing weight of 16,100 lb. the recommended final approach speeds in knots (m.p.h.) I.A.S. are:—

	Flaps down	Flaps up
Engine assisted	110 (125)	125 (144)*

*The aircraft requires a long landing run for a flapless landing and the throttles should be fully closed only after the aircraft has touched down. The initial straight approach should be made at a speed of about 20 knots (m.p.h.) above these figures.
Glide landings (see para 60).

48. Mislanding

(i) Open the throttles steadily to $+12$ lb./sq. in. boost and raise the undercarriage.

(ii) With the flaps fully down climb at 140 knots (160 m.p.h.) I.A.S.

(iii) Raise the flaps to the maximum lift position. At a safe height raise the flaps fully and retrim, there is no sink and the aircraft accelerates rapidly.

PART II—HANDLING

49. **After landing**

(i) When clear of the runway raise the flaps, set the radiator switches to OPEN, switch OFF the booster pumps and set the propeller control levers fully forward.

On reaching dispersal:—

(ii) On the last flight of the day check the ignition at zero boost (see para. 37). If the drop is excessive it should be noted in the Form 700.

(iii) Throttle back to 1,000—1,200 r.p.m., idle at this speed for about 20 seconds, and then stop the engines by operating the slow running cut-out controls.

(iv) When the engines have stopped, turn off the fuel, switch off the ignition and all electrical services, set the radiator flaps to AUTO and then switch the ground/flight switch to GROUND.

(v) Oil dilution

The correct dilution period for these engines is:—
One minute at air temp. down to $-10°C$.
Two minutes at air temp. below $-10°C$.

A.P. 2864C—P.N.
Pilot's Notes

PART III

OPERATING DATA

50. **Engine data Merlin 130 and 131**

(i) Fuel: 100 octane

(ii) Oil: D.E.D. 2472 (Stores Ref. 34A/115) Key letter Y

(iii) The principal engine limitations are as follows :—

		R.p.m.	Boost lb./sq. in.	Temp. °C. Coolant	Oil
TAKE-OFF	L	3,000	+18*	135	
MAX. CLIMB 1 HR. LIMIT	L ⎱ H ⎰	2,850	+12	125	90
MAX. CONTINUOUS	L H	2,650 ⎱ 2,850 ⎰	+9	105	90
COMBAT 5 MIN. LIMIT	L ⎱ H ⎰	3,000	+20*	135	105

*May not be used with less than 2,850 r.p.m.

OIL PRESSURE: Minimum in flight 45 lb. /sq. in.

MINIMUM TEMPERATURE FOR TAKE-OFF
OIL ... ... 15°C.
COOLANT ... ... 40°C.

A.L.I
PART III
para. 51
page 30

51. **Flying limitations**

(i) *Maximum speeds* in knots (m.p.h.) I.A.S. are :—
Diving
Below 5,000 ft. 345 (400)
5,000–10,000 ft. 328 (380)
10,000–20,000 ft. 293 (340)
20,000–30,000 ft. 241 (280)
Above 30,000 ft. 206 (240)
Undercarriage down 175 (200)
Flaps down 153 (180)

PART III—OPERATING DATA

A.L.I
PART III
page 31

(ii) *Maximum speeds* with 100 gal. drop tanks are :—

Sea level to 10,000 ft.	290 knots I.A.S.
10,000 ft. to 20,000 ft.	270 knots I.A.S.
20,000 ft. to 30,000 ft.	225 knots I.A.S.
Above 30,000 ft.	195 knots I.A.S.

The 100 gallon drop tanks may be jettisoned in straight and level flight at speeds between 150 to 260 knots I.A.S.

(iii) *Maximum weights*

For take-off and gentle manœuvres only	19,550 lb.
For all forms of flying	17,880 lb.
For landing	*16,100 lb.

* This weight represents a remaining fuel load of 350 gallons with all ammunition expended, or 280 gallons when landing with full ammunition.

Desert equipment must not be carried unless 25 gallons of fuselage fuel are omitted. (See para. 65 (ii)).

(iv) *Spinning*

Intentional spinning is prohibited. An incipient unintentional spin should be checked by normal recovery action at once.

(v) *Position error corrections*

From	220	240	260	280	300	320	340	knots
To	240	260	280	300	320	340	360	
Add	3	4	5	6	6	7	8	knots or m.p.h.
From	260	280	300	320	340	360	380	m.p.h.
To	280	300	320	340	360	380	400	

NOTE.—At speeds up to 220 knots and 260 m.p.h. the error is negligible.

PART III—OPERATING DATA

52. **Maximum performance and range**

    (*To be issued by amendment*)

PART III—OPERATING DATA

53. Fuel consumptions
(To be issued by amendment)

A.P. 2864C—P.N.
Pilot's Notes

PART IV
EMERGENCIES

54. Engine failure during take-off

(i) Safety speed with flaps up at typical service load (no external stores), 17,880 lb., is 145 knots (165 m.p.h.) I.A.S. at full take-off power, and is 135 knots (155 m.p.h.) I.A.S. at +12 lb./sq. in. boost and 3,000 r.p.m.

(ii)

(*To be issued by* **amendment**)

PART IV—EMERGENCIES

(iii) Taking off with the flaps at the maximum lift position lowers the safety speed by 3—5 knots.

(iv) If safety speed has been attained, the aircraft will climb away easily at the above speeds provided that the propeller of the failed engine is feathered.

55. Engine failure in flight

(i) If the failure is not immediately rectifiable, feather the propeller—See para. 56.

(ii) At typical service load, height can easily be maintained on either engine at moderate altitudes at about 165 knots (190 m.p.h.) I.A.S. at power well within the weak mixture range.

> NOTE.—In single engine flight, stability is considerably reduced especially in bumpy conditions. The application of yaw, either deliberately or accidentally, produces marked changes in longitudinal trim. Vibration of the rudder control may be experienced if full rudder is used.

56. Feathering

(i) Close the throttle of the failed engine.

(ii) Pull the propeller control lever through the feathering gate.

PART IV—EMERGENCIES

(iii) Press and release the feathering pushbutton (10), which should spring out when feathering is completed. If it does not do so it should be pulled out by hand.

(iv) Switch off the booster pump, turn off the fuel and switch off the ignition when the propeller has stopped, or nearly stopped rotating. When flying on the fuselage tanks, do not switch off the booster pump but only turn off the fuel cock of the failed engine. If the booster pump is switched off this may cause the live engine to cut.

(v) On early aircraft not embodying Mod. 338 there is only one generator, which is on the port engine. The capacity of the aircraft battery is small, and when the port engine is feathered, the drain on the battery should be reduced as much as possible by turning off all non-essential services.

57. Unfeathering in flight

(i) Set the throttle fully closed, the propeller control lever just forward of the feathering gate and the ignition on.

(ii) Turn on the fuel and press and release the feathering pushbutton which will spring out when r.p.m. rise to the minimum governing speed. Switch on the booster pump.

NOTE.—Do not unfeather at speeds greater than normal cruising speed owing to the danger of overspeeding.

58. Single-engine landing

(i) The single-engine performance of the aircraft is good and a left-hand circuit can safely be made (and is recommended) irrespective of which engine has failed.

PART IV—EMERGENCIES

(ii) Lower the undercarriage and the flaps to the maximum lift position as on a normal circuit maintaining a speed of 140 knots (160 m.p.h.) I.A.S. until the decision to land has been made. After which lower the flaps fully when required and make an engine assisted approach crossing the airfield boundary at the speeds quoted in para. 47.

A.L.I
PART IV
para. 59

59. Going round again on one engine

With the wheels down and the flaps at the maximum lift position, it is possible to go round again without loss of height, provided full power is used and the speed is not allowed to fall below 140 knots (160 m.p.h.) I.A.S.

---

60. Glide landing

Normally the aircraft should be landed from an engine-assisted approach as the rate of sink is very high with the flaps fully down and the power off. If it is necessary in an emergency to make a glide approach, it is recommended that the initial approach be made with the flaps in the maximum lift position, at a speed of 140 knots (160 m.p.h.) I.A.S. Lower the flaps fully just prior to levelling off using a final approach speed of about 130 knots (150 m.p.h.) I.A.S.

61. Undercarriage and flaps emergency operation

In the event of failure of the port engine (or of the hydraulic pump) there is sufficient pressure in the accumulator for one complete one-way operation of the undercarriage and flaps. If this fails, the flaps should be

PART IV—EMERGENCIES

A.L.I
PART IV
para. 61
page 38

lowered first by means of the handpump and then the undercarriage by setting the selector DOWN and pressing the toggle lever, giving it a quarter turn anti-clockwise and pulling.

If the flaps fail to come down with the handpump, or if they come down only partially and a leak in the system is therefore suspected, the flap selector must be returned to UP (to neutral if Mod 44—three position flaps—is not embodied) before the undercarriage is lowered. This prevents escape of compressed air through the flap system.

> NOTE.—The flaps should never be lowered by the compressed air system and if Mod. 339 is incorporated they cannot be.

62. Fire-extinguishers

Two engine fire warning lights (10) which incorporate dimmer screens are inset in the feathering pushbuttons. Two shielded red-painted fire-extinguisher pushbuttons are mounted on the pilot's electrical panel.

A.L.I
PART IV
para. 63

63. Hood jettisoning

The sliding hood can be jettisoned by means of the lever (13) marked HOOD JETTISON, mounted on the right-hand cockpit wall forward of the hood crank handle.

> NOTE.—On early aircraft not embodying Mod. 231 the hood crank handle must be given half a turn backwards before the jettison lever will release the hood.

64. Ditching

See A.P. 2095. No actual ditchings of this aircraft have been reported.

PART IV—EMERGENCIES

65. Emergency equipment

(i) A crowbar is stowed in spring clips on the floor of the cockpit to the right of the pilot's seat.

(ii) There is provision for the stowage of emergency desert equipment in the rear fuselage. Access from the outside is obtained by breaking in the portion of the fuselage marked DESERT EQUIPMENT.

PART V
ILLUSTRATIONS

	Fig.
Instrument Panel	1
Cockpit—Left hand side	2
Cockpit—Right hand side	3

INSTRUMENT PANEL

KEY TO *Fig.* 1
INSTRUMENT PANEL

1. Oil temperature gauges.
2. Coolant temperature gauges.
3. R.p.m. indicators.
4. Boost gauges
5. Oil pressure gauges
6. Fuel pressure warning lights.
7. Cockpit ventilator
8. Undercarriage indicator and warning light.
9. Flap position indicator.
10. Propeller feathering pushbuttons and fire warning lights.
11. Instrument flying panel.
12. Gyro compass indicator.
13. Hood jettison lever.
14. Oxygen regulator and flow indicator.
15. Windscreen de-icing regulator and handpump.
16. Fuel tank contents gauges.
17. Cross-feed cock switch.
18. Tank booster pump switches.

KEY TO *Fig.* 1

INSTRUMENT PANEL

1. Oil temperature gauges.
2. Coolant temperature gauges.
3. R.p.m. indicators.
4. Boost gauges
5. Oil pressure gauges
6. Fuel pressure warning lights.
7. Cockpit ventilator
8. Undercarriage indicator and warning light.
9. Flap position indicator.
10. Propeller feathering pushbuttons and fire warning lights.
11. Instrument flying panel.
12. Gyro compass indicator.
13. Hood jettison lever.
14. Oxygen regulator and flow indicator.
15. Windscreen de-icing regulator and handpump.
16. Fuel tank contents gauges.
17. Cross-feed cock switch.
18. Tank booster pump switches.

FIG 1

INSTRU

INSTRUMENT PANEL

FIG 1

KEY TO Fig. 2
COCKPIT
Left-hand side—seat removed

19. Undercarriage emergency (pneumatic) control.
20. Hydraulic handpump.
21. Main fuel (engine master) cock.
22. Heating control.
23. Bomb and drop tank manual emergency release.
24. Engine priming pumps.
25. Undercarriage selector lever.
26. Flap selector lever.
27. Supercharger gear change switch.
28. Throttle levers.
29. Propeller levers.
30. Pneumatic system and brake pressure gauges.
31. T.R. selector.
32. Elevator trim tab control and indicator.
33. Rudder trim tab control and indicator.

COCKPIT—LEFT HAND SIDE
—Seat Removed—

FIG 2

FIG 2 COCKPIT-LEFT HAND SIDE
— Seat Removed —

KEY TO *Fig.* 2

COCKPIT
Left-hand side—seat removed

19. Undercarriage emergency (pneumatic) control.
20. Hydraulic handpump.
21. Main fuel (engine master) cock.
22. Heating control.
23. Bomb and drop tank manual emergency release.
24. Engine priming pumps.
25. Undercarriage selector lever.
26. Flap selector lever.
27. Supercharger gear change switch.
28. Throttle levers.
29. Propeller levers.
30. Pneumatic system and brake pressure gauges.
31. T.R. selector.
32. Elevator trim tab control and indicator.
33. Rudder trim tab control and indicator.

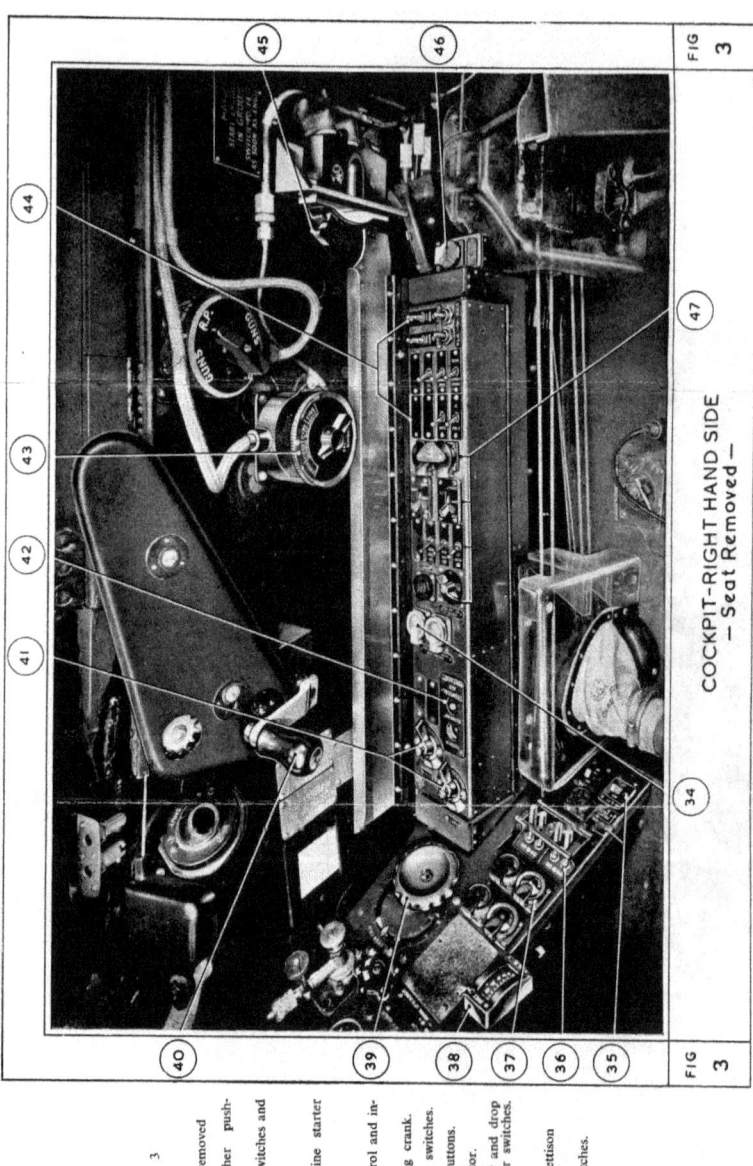

COCKPIT—RIGHT HAND SIDE
— Seat Removed —

KEY TO *Fig.* 3
COCKPIT
Right-hand side—seat removed

34. Engine fire-extinguisher pushbuttons.
35. Oil dilution selector switches and pushbuttons
36. Ignition switches.
37. Booster-coil and engine starter pushbuttons.
38. Suction gauge.
39. Aileron trim tab control and indicator.
40. Sliding hood operating crank.
41. Radiator flap control switches.
42. Air-intake filter pushbuttons.
43. Gyro gun sight selector.
44. Bomb fuzing, selector and drop tank jettison selector switches.
45. Ground-flight switch.
46. Bomb or drop tank jettison pushbutton
47. External lighting switches.

KEY TO *Fig.* 3

COCKPIT

Right-hand side—seat removed

34. Engine fire-extinguisher pushbuttons.
35. Oil dilution selector switches and pushbuttons
36. Ignition switches.
37. Booster-coil and engine starter pushbuttons.
38. Suction gauge.
39. Aileron trim tab control and indicator.
40. Sliding hood operating crank.
41. Radiator flap control switches.
42. Air-intake filter pushbuttons.
43. Gyro gun sight selector.
44. Bomb fuzing, selector and drop tank jettison selector switches.
45. Ground-flight switch.
46. Bomb or drop tank jettison pushbutton.
47. External lighting switches.

FIG 3 — COCKPIT-RI(— Seat

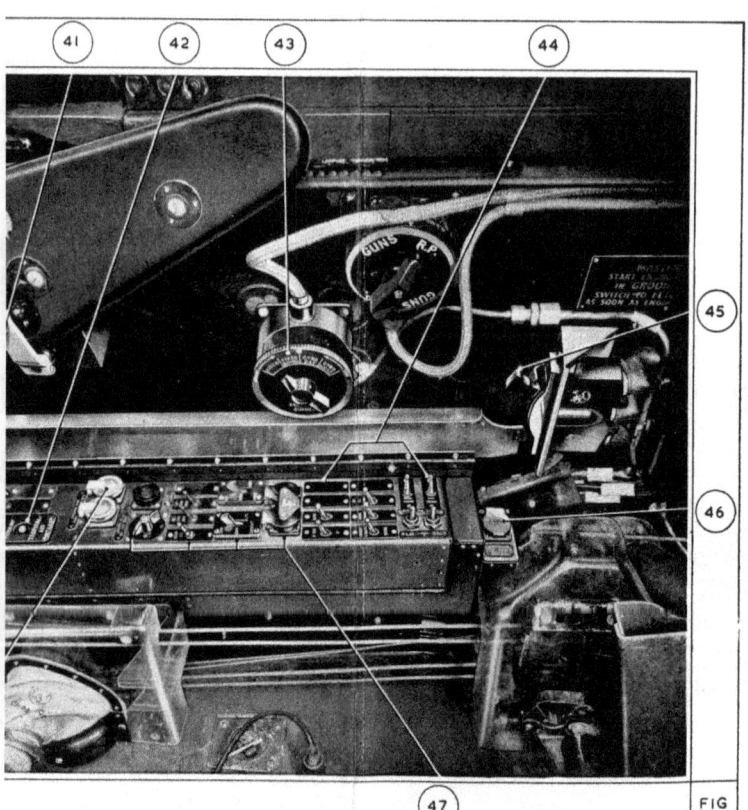

COCKPIT-RIGHT HAND SIDE
— Seat Removed —

FIG 3

FINAL CHECKS FOR LANDING

BRAKES ... OFF. CHECK PRESSURES

WHEELS ... DOWN AND LOCKED

PROPS. ... 2,850 RPM ON FINAL

FLAPS ... MAX. LIFT

FULL ON FINAL

FINAL CHECKS FOR TAKE-OFF

TRIM — ELEVATOR : NEUTRAL TO 1/2 DIV. NOSE HEAVY

RUDDER : NEUTRAL

AILERON : NEUTRAL

PROPS. — MAX. RPM

FUEL — MAIN COCKS OPEN

BOOSTER PUMPS ON

FLAPS — UP OR MAX. LIFT

©2012 Periscope Film LLC
All Rights Reserved
ISBN #978-1-937684-83-9
www.PeriscopeFilm.com